Be An Expert!

What Is the Weather?

Erin Kelly

Children's Press®
An imprint of Scholastic Inc.

Contents

It Is Sunny 4

It Is Cloudy 6

Know the Names

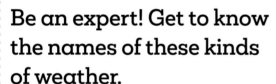

Be an expert! Get to know the names of these kinds of weather.

It Is Windy 12

A Rainbow! 18

It Is Raining 8

It Is Snowing 10

I See Lightning 14

I Hear Thunder 16

All the Weather 20

Expert Quiz 21

Expert Gear 22

Glossary 23

Index 24

It Is Sunny

The sun shines in the sky.
It is a beautiful day!

It Is Cloudy

Clouds fill the sky.
They hide the sun.

Be Prepared

Q: What is a cloud made of?

A: A cloud is made up of tiny **water droplets**.

It Is Raining

Water falls from the sky.
Grab your umbrella!

Expert Fact

When the water droplets in a cloud get big and heavy, they fall as rain.

It Is Snowing

Snow is falling.
It is cold today. Brr!

Be Prepared

Q: How do snowflakes form?

A: When the air gets cold, water droplets **freeze**. Snowflakes form as a lot of droplets freeze together.

It Is Windy

WHOOSH.
You can feel the air moving.
Wind can be gentle or strong.

Expert Fact

A **wind sock** can tell you in which direction the wind is blowing.

I See Lightning

FLASH!

The sky is lit up.

Be Prepared

Q: What do I do if I see lightning?

A: Go indoors. If you can't, stay away from trees. Lightning often **strikes** the tallest things in the area.

I Hear Thunder

BOOM!
Thunder is loud.

Expert Fact

Thunder is the sound lightning makes when it moves through the air.

A Rainbow!

Pretty colors appear in the sky. Can you name them all?

Be Prepared

Q: How is a rainbow formed?

A: A rainbow appears when the sun shines through water droplets in the air.

All the Weather

So many types of weather. What is the weather today?

Expert Quiz

Do you know the names of these kinds of weather events? Then you are an expert! See if someone else can name them too!

3.

4.

7.

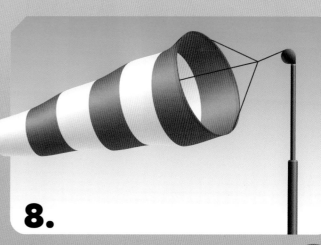

8.

Answers: 1. Cloudy 2. Rainbow 3. Thunder 4. Raining 5. Lightning 6. Sunny 7. Snowing 8. Windy.

Expert Gear

Meet a storm tracker. What gear does she need to study a storm?

She has a **flashlight**.

She has **radar**.

She has a **truck**.

She has a **raincoat**.

She has a **camera**.

22